SCIENCE AND SOCIETY™

MODERN GENETIC SCIENCE

NEW TECHNOLOGY, NEW DECISIONS

Terry L. Smith

ROSEN
PUBLISHING®

New York

Published in 2009 by The Rosen Publishing Group, Inc.
29 East 21st Street, New York, NY 10010

Copyright © 2009 by The Rosen Publishing Group, Inc.

First Edition

Library of Congress Cataloging-in-Publication Data

Smith, Terry L. (Terry Lane), 1944–
Modern genetic science : new technology, new decisions / Terry L. Smith.
    p. cm.—(Science and society)
Includes bibliographical references and index.
ISBN-13: 978-1-4358-5027-9 (library binding)
1. Genetic engineering. 2. Genetics. I. Title.
QH442.S575 2009
660.6'5—dc22

                                    2008013418

Manufactured in Malaysia

On the cover: A lab technician uses monoclonal antibodies to detect a bacterium. Monoclonal antibodies are produced by a single clone of cells, which can then be made in large amounts in the lab to help in targeting specific substances that are recognized by the immune system.

CONTENTS

"**I**sn't that a Triceratops?" "There goes a T. rex!" What fun—opening day at a new amusement park where dinosaurs roam free! The scientists who brought the dinosaurs back to life from preserved DNA are eager to see the success of their experiment. The investors who paid for it hope to make lots of money. The first visitors can hardly wait to see the dinosaurs that roam through the forest. But suddenly, everything goes very wrong. A computer failure knocks out the electronic devices that keep the animals under control. The dinosaurs go on a rampage, injuring some of the visitors. Much excitement follows, as officials try to round up the dinosaurs and visitors hurry to escape.

Yes, Jurassic Park was only a movie amusement park. When the excitement of the movie *Jurassic Park* (1993) ends, everyone can relax, knowing that such a notion could never really happen. Scientists assure the public that there is no way they will ever see a living dinosaur. But short of dinosaurs, what can people expect now that scientists are able to create new forms of life by controlling the genetic code? Maybe cats that don't cause allergies? Or plants with black leaves so they can absorb more sunlight? What about grass that never needs mowing? Or bacteria that can tell if a terrorist is carrying explosives?

Just like in the movie *Jurassic Park*, a lot of people are excited about the benefits that genetic engineering can bring. It helps doctors diagnose and treat diseases. Farmers can grow crops more easily, thanks to modern genetic science. It is helping to make your world a safer and cleaner place in

The movie *Jurassic Park*, from which this film still was taken, is about scientists who clone dinosaurs for a theme park. In the movie, to clone the dinosaurs, scientists used dinosaur DNA from some blood they had extracted from an ancient insect that was encased in amber. In reality, dinosaurs have been extinct for millions of years, and none of their genetic material has been preserved.

which to live. Someday, it may even provide you with a clean and plentiful source of fuel. However, just as in the movie *Jurassic Park*, people need to be concerned about the consequences of genetic engineering that they might not expect.

Scientists are only beginning to learn what can be accomplished through genetic engineering. There may be unknown risks. Moreover, because the technology is so powerful, a failure of some sort could be disastrous. Many people warn that scientists should proceed slowly until they are certain a new application is safe. Beyond safety, some procedures that can be done using

A sample of cells from a plant or animal can be analyzed to determine the organism's genetic code. The process is complex, but modern technology makes it quite fast and easy.

genetic engineering may be unethical. For example, most everyone agrees that certain measures, such as making cloned humans, should never be attempted or done.

What is genetic engineering? Very simply, the term means "building with genes." Inside every cell, genetic material, in the form of DNA, tells that cell what to do. Certain bits of DNA, called genes, play special roles. A gene in a tomato plant's DNA makes the tomato red, poodles have genes for curly hair, a bacterial gene produces a chemical that causes food poisoning, and so on. Scientists know about thousands of genes, in all kinds of organisms. By experimenting, they can figure out what each gene does. A gene's DNA can be moved about, using specialized methods. A gene can be cut out from one cell and put into another. The process of moving bits of DNA to create new organisms is referred to as genetic engineering.

"Biotechnology" is a term related to genetic engineering. It refers to the methods applied to make products using genetic engineering, or to the companies that manufacture these products. Another process related to genetic engineering is called cloning. Cloning generally involves moving the entire genetic code from one cell to another, rather than moving only small pieces of DNA.

You live in the Age of Biology. Your world is changing quickly, whether you like it or not. Many of these changes are brought about by genetic engineering. Some people owe their health to medicines produced by genetic engineering. Others worry that genetically engineered plants or animals may cause harm. Whatever the concerns are, knowing more about genetic engineering will help everyone to participate in the fast-changing world.

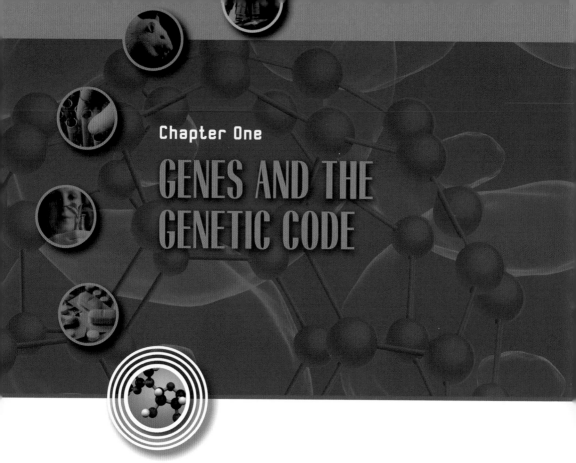

Chapter One

GENES AND THE GENETIC CODE

What a variety of organisms share the earth, from tiny bacteria to huge elephants. Yet, every plant and animal begins life as a single cell packed with instructions telling it to grow into a tall oak tree, a slithery snake, or whatever it is supposed to be. In the case of human beings, that single cell gives rise to the trillions of cells in a mature human. During the division process, these cells are taking on specific functions. Some become bone cells, others become brain cells, and so forth, until finally there are more than two hundred types of cells that make up your body. Together, these cells carry out the complex tasks required for your body to function. How is it possible that the information to bring about all these complex changes is packed into that first tiny cell?

The Genetic Code

Cells are much too small to see without a powerful microscope, but if you could look inside one of your cells, you'd see a structure called a nucleus. Packed within that nucleus are tiny strands called chromosomes. These chromosomes contain one's complete genetic code. Each organism has a unique number of chromosomes. Human cells have forty-six.

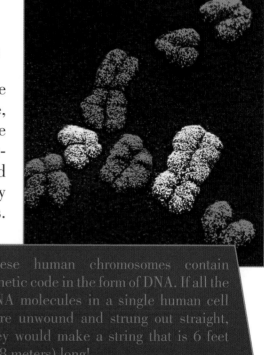

These human chromosomes contain genetic code in the form of DNA. If all the DNA molecules in a single human cell were unwound and strung out straight, they would make a string that is 6 feet (1.8 meters) long!

Chromosomes are made up of a chemical called deoxyribonucleic acid, or DNA for short. The chemical structure of DNA looks similar to a ladder that has been twisted into a spiral shape. It is often referred to as a double helix. This unusual shape allows it to be packed tightly into the cell's nucleus. The "rungs" on the DNA ladder are pairs of chemical units called bases, and there are only four different kinds: A (for adenine), C (for cytosine), T (for thymine), and G (for guanine). Amazingly, these four bases that make up the rungs of the DNA ladder act as a sort of alphabet that spells out all the instructions in a person's genetic code.

How is it possible that all the instructions to form and operate your body are spelled out with an alphabet consisting of only four letters? The secret is that your DNA has so many of these bases, or "letters" in the "alphabet," repeated in long series. Scientists estimate that there are about three billion of them, and all together they make up what is called the human genome. Genes are series of these bases along the DNA "ladder" that play special roles in telling the cells what to do. Genes are able to exert their effects on the body by instructing cells to make proteins. The body's thousands of types of proteins provide structure to the cells and carry out the actual functions of life.

Genes are arranged on the chromosomes in a precise order. In all, there are an estimated twenty thousand to twenty-five thousand genes in a human cell. This is not surprising when you consider the complexity of the human body.

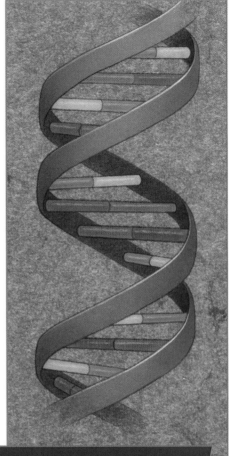

The genetic code of an individual is determined by the order of base pairs. These are tightly packed within the double helix structure of a DNA molecule.

Heredity

Your body's cells are dividing constantly throughout your life. Usually, when a cell divides into two cells, all the DNA in the cell nucleus is duplicated. Each resulting cell has forty-six chromosomes and identical DNA. When cells divide to form sex cells, however, the process is different. In this case, each resulting cell has only twenty-three chromosomes and half the amount of DNA. When an egg cell from the mother is fertilized by a sperm cell from the father, the DNA from the two cells is joined. In this way, a person's DNA is inherited from his or her two biological parents. Each time this process occurs, the DNA from the mother and the father mixes together in a different way. This is why brothers and sisters may appear very different, even though they have the same parents. The exception is identical twins, who inherit essentially identical DNA from their parents. This same process of inheritance occurs in plants and animals.

Engineering Genes

To engineer something as tiny as a gene, scientists first apply what they have learned about the DNA that makes up the genes they wish to change. Then, they use tools and procedures that allow them to manipulate genes to accomplish their goal.

For instance, a genetic engineer has identified a gene in one organism that he or she wishes to insert into the DNA of another organism. A kind of genetic "scissors" is first used to snip out the gene from its location in the first organism's DNA. These so-called scissors are actually chemicals called restriction enzymes. The scientist chooses an enzyme that will snip the DNA in the precise location of the gene of interest. The same enzyme is used to cut the second organism's DNA. These pieces of DNA are mixed together in the presence of a chemical "glue." Their loose ends combine, forming a new DNA combination. This new DNA is referred to as recombinant DNA, or rDNA for short.

The making of recombinant DNA is key to the many forms of genetic engineering. Because the genetic code of all organisms is made up of DNA with a similar chemical structure, it is possible to transfer genes from one organism to another. Often, a gene from another organism is transferred into the genetic material of

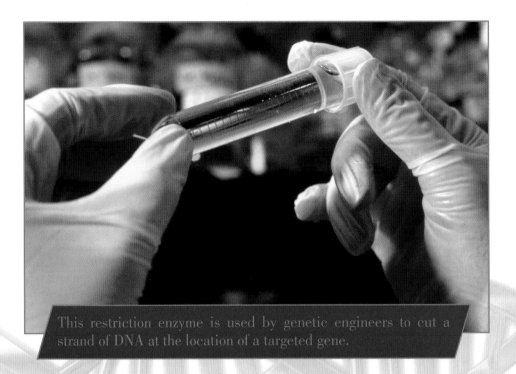

This restriction enzyme is used by genetic engineers to cut a strand of DNA at the location of a targeted gene.

bacteria. Bacteria have the ability to multiply rapidly. Sometimes, scientists take advantage of this characteristic by inserting a gene that instructs for production of a needed protein. As the bacteria multiply, the inserted gene causes the bacteria to produce more and more of the protein. Or viruses may be used to carry a gene into the cell of another organism, acting as a sort of genetic engineer's assistant. Viruses are tiny particles with the ability to invade cells. Once inside, they hijack the cell's genetic material in order to reproduce themselves. Scientists take advantage of this ability by splicing normal human genes into a virus. The virus can then enter a diseased human cell and restore it to a normal state.

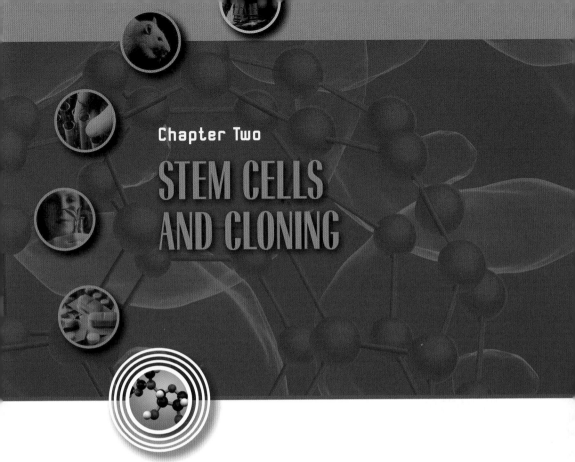

Chapter Two

STEM CELLS AND CLONING

Specialists are people with particular training or abilities to carry out some difficult task. Similarly, most cells of the body have been programmed by their DNA to carry out specialized needs. They may be muscle cells, skin cells, nerve cells, or liver cells. Stem cells, though, are different—they are not specialized to do a specific function. Instead, they have the very important ability to turn into specialized cells when they are needed. They have one other important characteristic not shared by other cells. They are able to divide and make new stem cells for a very long time. These attributes make stem cells very useful to geneticists.

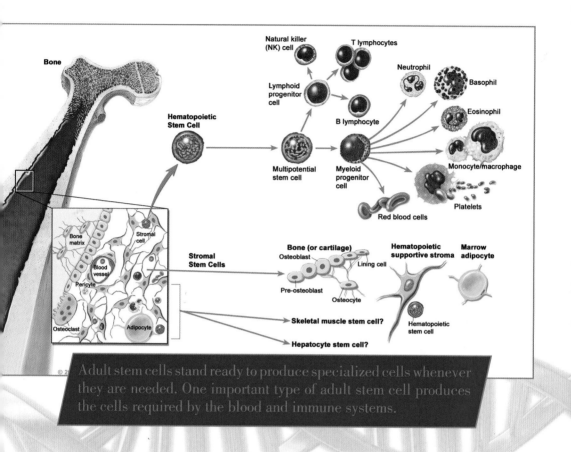

Adult stem cells stand ready to produce specialized cells whenever they are needed. One important type of adult stem cell produces the cells required by the blood and immune systems.

Stem Cells and Reproduction

The joining of female and male sex cells is the first step in animal reproduction. After a few days, this first cell has divided to produce a small clump of cells. These cells are embryonic stem cells. They will grow to form an embryo, an early stage in reproduction. As the cells continue to divide, some of them change into specialized forms, until finally a complete organism with dozens of cell types is produced. Each new cell has the same DNA as the original stem cells, but different genes are "turned on" or "turned off" as the cells take on their specialized roles. For example, those cells that become red blood cells have turned on the gene that allows them to carry oxygen to other tissues. Some stem cells remain, even in adult organisms. These are called adult stem cells. These are able to produce only a limited number of cell types, not a complete organism. For example, blood-forming stem cells are able to produce all the types of blood cells in the body.

Stem Cells in the Laboratory

If human embryonic stem cells can grow into any other type of cell, then why not use them to repair body parts when they are needed? This application is exactly what scientists in the health field are working toward. For example, if someone's heart muscle is damaged, it might be possible to give that person new heart cells grown from stem cells. Before scientists can apply these new methods, they need to do further research using stem cells. This research, though, has become controversial because of where the stem cells come from.

Infertility clinics help couples who are not able to have children. One method involves producing embryos in a laboratory from sex cells obtained from the mother and father. For various reasons, these embryos may be rejected by the couple responsible for them. In other words, the embryos in question will never be put into a woman's womb to grow into a human being. Some people think scientists should be allowed to use these embryos for stem cell research. They believe this is justified because by allowing scientists to study them, it is likely that many people will be helped. Critics of stem cell research believe it is wrong to experiment with embryonic stem cells because those cells have the ability to become a human being. The controversy was so strong that U.S. government funding for stem cell research was restricted by presidential order in 2001.

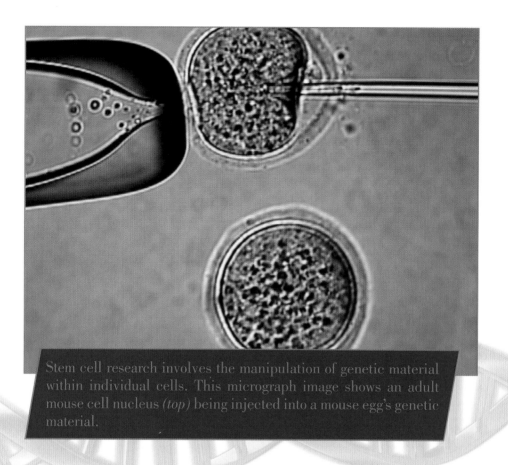

Stem cell research involves the manipulation of genetic material within individual cells. This micrograph image shows an adult mouse cell nucleus *(top)* being injected into a mouse egg's genetic material.

An exciting discovery was announced in 2007 that may put an end to this difficult ethical controversy. Two groups of scientists working independently in Wisconsin and Japan announced that they had produced human embryonic stem cells from skin cells. They did this by using genetic engineering methods to insert four new genes into the DNA of some skin cells. They knew that these particular genes were very powerful and that they had the ability to turn many other genes on or off. The addition of these four new genes gave the skin cells all the characteristics of embryonic stem cells. Embryos would no longer be required for scientists to experiment with embryonic stem cells.

What Is Cloning?

Would you want to go to a shopping mall and see dozens of people who look exactly like you? Surely not. Although cloning would make it possible to produce any number of identical persons, almost everyone agrees that cloning humans would be unethical and that it should never be done. However, many believe it is OK to reproduce animals by cloning.

A clone is an exact copy of an animal. The original animal and clone share identical genetic material. Mostly, they look and act alike. You might think only one animal would be required to produce a clone, but gen-

Dolly was the world's first cloned sheep. Animal cloning may help to preserve endangered species and improve food production.

erally the process requires three. An egg cell is obtained from one animal, a cell from a second animal provides the DNA, and a third animal provides the womb. Genetic engineering tools are used to remove the nucleus, along with its DNA, from the first animal's egg cell. Then, the nucleus and DNA from a cell of the second animal, the one being "copied," is placed in the egg cell. The resulting cell acts like a fertilized egg cell when placed inside the womb of yet a third animal. When born, the young animal is a clone of the second animal because that is the source of its DNA. Another term for cloning is "nuclear transfer."

Cloned Javan Banteng

The banteng, a relative of the cow, can be recognized by its white stockings and slender, curved horns. Only a few thousand remain in the wild on the island of Java. Scientists working with the San Diego Zoo in California were able to clone a banteng using genetic material preserved from a banteng that had died twenty years earlier. The preserved DNA was put into an egg cell from a dairy cow, and the resulting embryo was brought to maturity in the womb of a dairy cow. The resulting banteng can be seen at the San Diego Zoo. Visitors to the zoo are reminded that, despite success with the banteng, cloning alone cannot save endangered species. It is also essential to protect the habitats of these species.

The Javan banteng is an endangered species. Scientists made use of cloning technology to produce this male Javan banteng at the San Diego Zoo.

The first and most famous cloned mammal was a sheep, named Dolly, born in Scotland in 1996. Since then, many other animals have been cloned. It is a difficult process, and only a few attempts result in a healthy animal. Some cloned animals have died unexpectedly. Dolly developed cancer and died while still fairly young. Still, scientists think it is important to be able to clone animals, especially animals that have special qualities. For example, some animal species are in danger of becoming extinct. If scientists can preserve the DNA from some of the last remaining animals, then it may be possible to create new ones in the laboratory. The giant panda and the Sumatran tiger might eventually be saved from extinction using cloning methods.

Sometimes, the object of cloning experiments is not to produce a cloned animal. Instead, scientists use cloning techniques to make stem cells. The hope is that these stem cells can be used to produce cells, tissues, or organs to repair or replace diseased human body parts. This type of cloning is sometimes referred to as therapeutic cloning. Because this type of cloning research has the aim of improving human health, rather than reproducing a cloned human, some cloning critics find it acceptable.

Chapter Three

FEELING SICK? GENES TO THE RESCUE

A visit to the doctor's office is usually very predictable. After many questions and a few tests, the doctor very likely writes the patient a prescription for a common medicine. But imagine a doctor's office of the future. Perhaps one of the first things a doctor will want to know about a patient is what's in his or her DNA. It could help decide exactly what's wrong with the patient. The doctor would choose a treatment designed especially to match that patient's illness. Most likely the medicine would have been made using genetic engineering methods. The doctor might recommend that brothers and sisters have their DNA tested to learn if they might come down with the same disease.

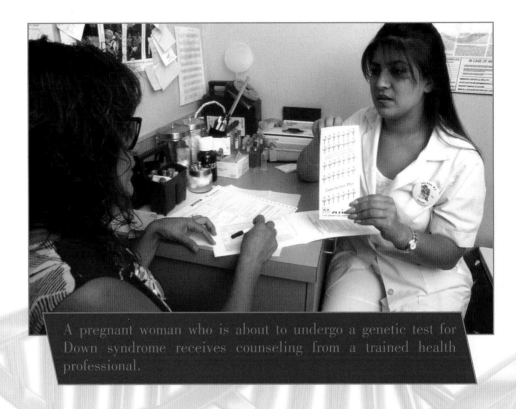

A pregnant woman who is about to undergo a genetic test for Down syndrome receives counseling from a trained health professional.

In fact, some of these methods are already available. Genetic testing is in wide use for certain inherited diseases. Genetic engineering methods are used to produce dozens of commonly prescribed medicines. But modern medicine can also be expensive. Will everyone who needs it be able to afford it? Are the methods used to achieve these medical advances ethical? These are difficult issues that our society will have to decide.

Genetic Testing

Everyone has a unique set of genes, inherited from parents. Once in a while, certain genes can be defective, leading to genetic disease. Even if parents are healthy, it is possible that their genetic code may carry defective genes that could lead to genetic disease

A researcher microinjects DNA into a cell to modify genetic material. Genetic research will play a major role in medical treatments of the future.

in a child. There are thousands of diseases related to defects in genes, but most of them are rare. Some of the more common ones with a direct genetic link are cystic fibrosis, hemophilia, and sickle cell disease.

Genetic testing involves analyzing a small sample of blood or tissue. If a patient is suspected of having a genetic disease, then doctors will examine a DNA sample for a defect in the gene corresponding to that disease. The results may help doctors decide the proper diagnosis and treatment for the disease.

Sometimes, genetic testing is done on healthy people. If a person has a relative with a genetic disease, then it is possible that he or she has inherited the same defective gene. Is it good to know that one has inherited a gene that could cause a disease? That depends on the gene and the disease. Some genes have high penetrance. This means a person is almost certain to get the disease if that form of the gene is inherited. Many genes have low penetrance. If a person inherits a gene for a disease with low penetrance, then he or she may or may not get the disease. That person just has a bit higher chance of getting it. For example, some genes make it more likely that a woman will get breast cancer. If a woman knows she has one of these genes, then she might take steps to reduce her breast cancer risk. On the other hand, knowing she has the breast cancer gene might cause a lot of worry. Most agree that the results of genetic testing should be private. In the United States, the Genetic Information Nondiscrimination Act of 2008 protects individuals from discrimination by employers or health insurers based on their genetic information.

Making New Medicines

Insulin is essential to regulate the amount of sugar in the bloodstream. Persons with diabetes are not able to make enough insulin in their own bodies and, therefore, must receive it regularly in the form of medicine. Drug companies produce insulin using recombinant DNA technology. The gene with instructions for making insulin is snipped out of a normal cell using the appropriate DNA "scissors." It is then inserted into a common form of bacteria. The bacteria multiply rapidly. Soon, lots of bacteria, all with

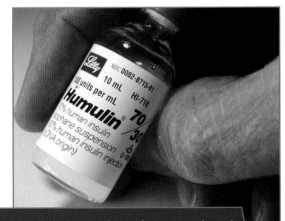

A synthetic form of insulin is produced using recombinant DNA technology. Millions of people with diabetes depend on it for their well-being.

the inserted gene, are producing insulin. It can then be removed, purified, and packaged for use by diabetic patients.

Researchers are at work every day using genetic engineering methods to discover new medicines to treat disease. Many of these are for the treatment of cancer. Cancer is not just a single disease but many different diseases. Doctors have learned to test the DNA of cancer cells to learn exactly what type of cancer a patient has. At the same time, medicines have been developed that are able to fight cancer cells with certain defects in their DNA. Eventually, doctors may be able to treat each cancer patient with medicines customized for his or her form of cancer.

Using Stem Cells

The repair of heart muscle using cardiac stem cells? New skin developed from hair roots? Stem cells used to cure some forms of cancer? Replacement blood cells used to treat sickle cell anemia? Today, these and many other treatments are being worked on in research laboratories around the world. The ability of stem cells to change into other types of cells makes them extremely valuable for treating certain diseases. They may be able to generate new tissue, or even entire organs, and provide a cure for diseases with few other treatment options. Although it may take many years to perfect the procedures, scientists have already made good progress. Someday, stem cells could become routine treatment for diseases like spinal cord injury, Parkinson's disease, and stroke.

Gene Therapy

Perhaps you're thinking, "If doctors can find a gene that is causing a disease, why don't they just fix the gene?" Unfortunately, it is not simple to change genes within the human body. For inherited diseases, the same defective gene is present in every one of the body's trillions of cells. A few gene treatments have been tried, using viruses to carry normal genes into the cells with defective

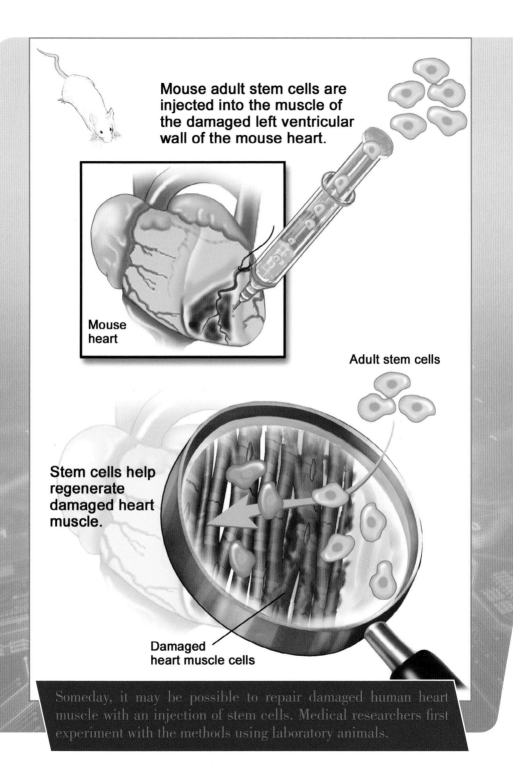

Mouse adult stem cells are injected into the muscle of the damaged left ventricular wall of the mouse heart.

Mouse heart

Adult stem cells

Stem cells help regenerate damaged heart muscle.

Damaged heart muscle cells

Someday, it may be possible to repair damaged human heart muscle with an injection of stem cells. Medical researchers first experiment with the methods using laboratory animals.

genes. These have not been successful, and research continues on new ways to do gene therapy. Another difficulty is that many genetic diseases are the result of defects in more than one gene. Scientists would have to know all the genes that cause a disease and then find a way to correct all these genes. Clinical trials of several types of gene therapy are underway at major medical centers. Some of the diseases for which gene therapy is being attempted are muscular dystrophy, Parkinson's disease, diabetes, arthritis, immune deficiency diseases, and many forms of cancer. These trials are in very early stages of research, but they hold the promise of effective new treatments for many common diseases.

Chapter Four

ANIMALS AND GENETIC ENGINEERING

L ong before anyone ever heard of genes, farmers were "engineering" the genes of their farm animals by selecting their best animals for breeding purposes. Today, genetic methods are available to "engineer" animals to make them more useful for human purposes.

Selective Breeding

Once in a while during the natural process of cell division, a mistake occurs in the copying of a gene's DNA. The resulting gene is called a mutation. If the mistake occurs in a sex cell, then the mutated gene may be passed on to future generations. Occasionally, a mutated gene will result in an animal with a special

This thoroughbred colt has inherited desirable characteristics from its mother and father. Genetic engineering methods allow scientists to control an animal's inherited characteristics.

characteristic. If this characteristic gives a racehorse the ability to run fast, then that animal is more likely to be used for breeding purposes. This process of selecting animals with the best characteristics to serve as parents to the next generation of animals is called selective breeding. Although selective breeding relies on naturally occurring genetic events, scientists are now able to produce changes in genes using genetic engineering.

Transgenic Animals

How about a glowing pet fish to impress your friends? Since 2003, it has been possible to purchase the world's first transgenic

Transgenic animals, such as these glowing zebra fish, are helping scientists to make medical discoveries, clean up pollution, and manufacture useful chemicals.

pet—a zebra fish that has been genetically engineered to glow in the dark. To create the fish, scientists identified a gene from a jellyfish that caused the jellyfish to glow. They used their genetic "scissors" to remove this gene from the DNA of a jellyfish. Next, they inserted the gene into the DNA of a zebra fish. The resulting fish glows the same as the jellyfish. The glowing zebra fish is an example of a transgenic animal.

A transgenic animal is one whose DNA is a mixture of DNA from two organisms. The glowing zebra fish was initially developed because scientists were looking for a method to detect water pollution. Their intention was to insert a gene into the DNA of the zebra fish that would cause it to glow if certain chemicals were present in the water. Then, they noticed that the fish were also fun to look at. Most experiments with transgenic animals involve changing an animal to make it more useful to humans. Some people think it is not ethical to interfere with the natural creation of animals, even if the purpose is to benefit society.

Use of Animals in Medical Research

Scientists make discoveries by experimenting. But what if a medical scientist is trying to learn about human genes? He or she cannot experiment with human beings. Instead, scientists take advantage of the fact that animals have many of the same genes as humans do. This is why mice are so important to genetic scientists. Mice have many genes that correspond to human genes. Besides, they are small, they reproduce rapidly, and they do well under laboratory conditions. To study a certain human gene,

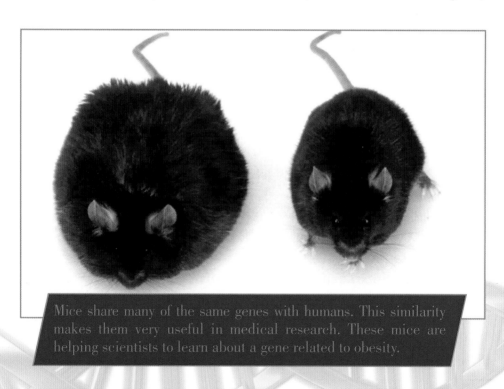

Mice share many of the same genes with humans. This similarity makes them very useful in medical research. These mice are helping scientists to learn about a gene related to obesity.

scientists look for that same gene in the mouse genome. Then, they can experiment on mice to find what happens when that gene is changed, using the tools of genetic engineering. The results help them to understand more about human genes.

One important type of laboratory mouse is called a knockout mouse. Researchers insert an artificial piece of DNA in a mouse's genome in place of some gene they want to study. This procedure makes the gene inactive, or "knocks it out." The researchers then observe how the mouse differs from a normal mouse. These differences help explain the function of the gene being studied. There are many types of knockout mice, depending on what gene has been inactivated. One type of knockout mouse is helping doctors learn about asthma. A certain human gene is suspected of making a person more or less likely to get asthma. Mice have this same gene. Scientists created knockout mice in which the suspected gene was inactivated. These mice developed symptoms of human asthma. The results proved that the scientists' suspicions about the human asthma gene were correct.

Products from Animals

Animals have long served as a source of food and clothing for humans. Today, through genetic engineering, animals are providing many products for human use. Animal cells are capable of producing thousands of kinds of proteins, according to the instructions of their DNA. Many of these proteins are useful for treating human diseases. By adding genes into an animal's DNA, scientists can develop an animal that produces a certain protein in its milk. A goat farm in Massachusetts raises transgenic goats that produce in their milk a protein used in treating human blood diseases. A Dutch company is using cows to produce a protein that boosts the human immune system. A company in Canada has developed transgenic goats with a gene for the production of spider silk that is so strong that it can be used in bulletproof

Looking for the Fountain of Youth

No more wrinkles! No more gray hair! Scientist Cynthia Kenyon is on a mission. She wants to keep people from growing old. But she's starting with a tiny worm called a roundworm. She has discovered a gene, which she calls daf-2, that controls the worm's aging process. By turning off this gene, Dr. Kenyon has been able to grow worms that live six times longer than normal. Moreover, they not only live longer, but they also stay sleek and squirmy, just like young worms.

Other scientists are also on the lookout for the "aging" gene. But finding a way to apply these methods to humans is probably a long time away. Even so, some scientists and scholars think it is time to slow down and consider the consequences. Would antiaging treatments have undesirable effects or perhaps prolong years of ill health? Would our society be able to support a huge increase in the number of old people?

vests. Animals at these farms are well cared for and healthy. Still, there are concerns by some that the genetic changes might eventually result in disease, or that it is unfair to the animals to alter their genes for human purposes.

Cloning of Animals

A calf bred from a prize dairy cow might or might not inherit her mother's gene for high milk production. But if that prize cow were cloned, the calf would be certain to inherit good milk-producing genes. The new calf would be born with DNA identical to the cloned animal. Similarly, cattle ranchers would like to reproduce those cows that grow fastest and produce the most beef. According to the U.S. Food and Drug Administration (FDA), milk and meat from cloned animals or their offspring are safe to eat. Some people do not like the idea of eating food that is not naturally produced. Others think cloning of agricultural animals is good because it allows farmers to produce more food at a lower cost. Labeling meat and milk that comes from cloned animals or their offspring would let consumers decide for themselves whether they want to purchase these foods or not.

If your favorite pet died, would you want scientists to produce a new puppy that was a clone of it? Some people say yes, but most say no after they think about it for a while. For one thing, cloning techniques are not yet perfect, and the puppy might be sickly or might die at a young age. Differences in diet might cause the puppy to be not exactly the same as the favorite pet. Cloning would also take away the excitement of discovering all the new things about a puppy that is born the old-fashioned way.

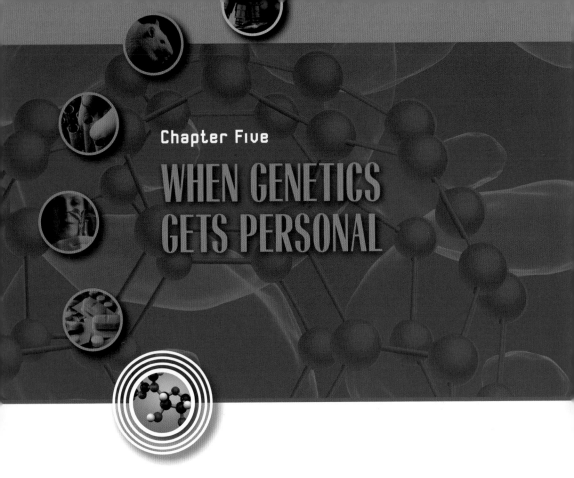

Chapter Five

WHEN GENETICS GETS PERSONAL

I t is an exciting step when a young couple decides it is time to start a family. It is also a bit scary. Every expectant parent worries about what the new baby will be like. Will it be healthy? Will it be tall like its grandfather, have red hair like its mother? Parents hope for a perfect child. But just what makes a perfect child, and who decides? Is it important for people to know about the genes they've inherited from their parents? Issues about one's personal genetics are controversial—and also hard to ignore.

Pregnancy Genetic Testing

It is now a routine procedure for doctors to take a sample from the womb of a pregnant woman for

A lab worker prepares samples for genetic testing. Samples taken from a pregnant woman can determine if her baby will be born with an inherited disease or malformation.

DNA testing. The sample can then be analyzed to determine if the baby will be born with a serious disease due to genetic defects. If so, then the parents may choose not to continue the pregnancy. They may think it is unfair to bring a child into the world who will not be able to lead a normal life. Others think it is unethical to destroy something that has the potential to become a human being. The idea of pregnancy testing for disabilities is offensive to many people with inherited disabilities. To them, it may suggest that they are not valuable human beings.

Designer Babies

Couples can now use genetic techniques to choose the set of genes their new baby inherits. Many babies born today get their start in a laboratory, through in vitro fertilization. This procedure results

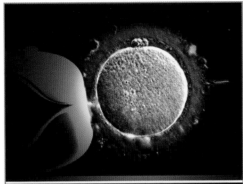

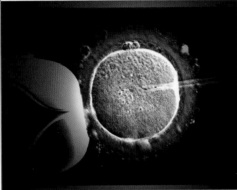

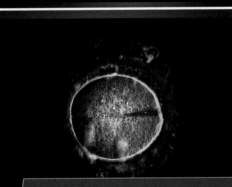

These photos show injection of a sperm cell into an egg cell during in vitro fertilization. The resulting embryo can be screened for genetic characteristics before being placed in the mother's womb.

in the production of many fertilized eggs. After a few days of cell division, a tiny ball of cells from a fertilized egg is placed in the mother's womb, where it will mature into a baby. Which of the many fertilized eggs is chosen? Can the parents choose a "perfect" baby?

Perhaps the family is affected by a serious genetic disease. The DNA of each fertilized egg can be examined to see if it carries the genetic defect that causes the disease. An egg without the disease gene is chosen to be placed in the mother's womb. The parents could look forward to having a healthy baby.

It is a different case when parents are not worried about a genetic disease. Instead, parents might want to choose a baby based on another characteristic, like whether it will be male or female, or whether it will have a blood type that matches that of a sick brother or sister. In other words, they want a "designer baby."

In the future, parents might use genetics to choose any number of traits they would like their child to have.

A major medical group, the American Society for Reproductive Medicine, opposes the common use of genetic information to choose babies, except for prevention of inherited disease. However, other doctors support parents' rights to choose a baby's sex or other characteristics. Genetic techniques are becoming cheaper and more available. This accessibility presents difficult decisions for many young couples.

Knowing Your Genetic Code

Some people need to know about their genetic inheritances for health reasons. But much genetic testing these days is done for less serious reasons. Now, anyone can find out if they have a certain mutant gene, or find out their entire genetic code, just by mailing in a test sample in response to an advertisement on the Internet. Genetics laboratories claim to provide all kinds of information based on a DNA sample, such as what kinds of foods one should eat, who one's ancestors were, or whether one is likely to be addicted to nicotine.

How should you respond to an advertisement for having your personal DNA tested? Very cautiously, according to genetics experts. The American College of Medical Genetics advises consumers to involve a genetics expert in the process of genetic testing. There are few regulations to make sure that DNA testing laboratories provide accurate results. Even if results are accurate, most individuals are not qualified to understand what they really mean. The results might cause unnecessary worry if a person thinks he or she will develop a deadly disease. In those cases when a doctor recommends a particular genetic test, persons trained in genetic counseling are available to help an individual understand and deal with the results.

MYTHS AND FACTS

MYTH Cloned animals look and act exactly alike.

FACT. Although cloned animals share identical genetic material, there may be small differences in appearance. For example, cloned cows may have different patterns of spots or shapes of ears. Behavior and temperament are influenced by life experiences as well as by genetic inheritance, so cloned animals may behave differently.

MYTH KFC no longer has the word "chicken" in its name because it doesn't serve chicken.

FACT. A rumor was circulated that scientists had genetically engineered a chicken to do away with the inedible parts, such as feathers and beaks. The rumor claimed that the animal was so changed that KFC could no longer refer to it as chicken. The truth is that the name was changed for commercial reasons, and that KFC serves the same chicken that it always did.

MYTH Genetic tests are regulated by the government and are totally reliable.

FACT. Laboratories that provide genetic testing are not inspected or regulated by any government agency. Many in the scientific community are concerned that laboratories may be providing incorrect results to consumers. Furthermore, test results, even if they are correct, may not be adequately explained so that a consumer can understand them.

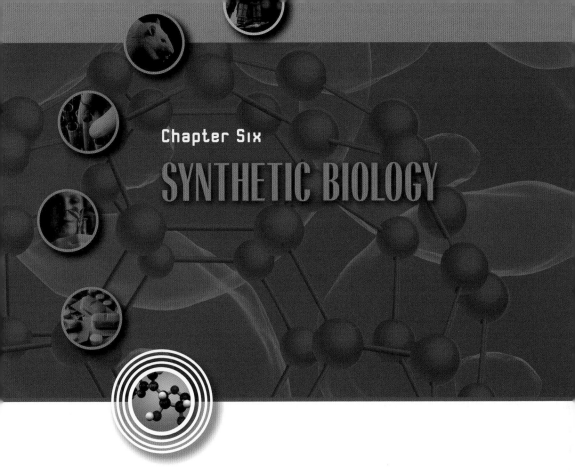

Chapter Six

SYNTHETIC BIOLOGY

T he five members of the team from Mr. Cachianes's high school biotechnology class participated in the finals of the International Genetically Engineered Machine competition in Cambridge, Massachusetts. The competitors were more than fifty college-level teams from around the world. The high school students had spent an intense summer building their contest entry—yeast cells in which they had "built" a tiny new organ— in a laboratory at the University of California, San Francisco.

The object of the competition was to build a simple biological system from standard parts. The students received a kit containing various bits of

These nine biotechnology students from Abraham Lincoln High School in San Francisco, California, beat out several college-level teams with their entry in a contest for the new scientific field of synthetic biology.

DNA—similar in idea to a set of genetic toy building blocks. They could use these in building their entry. In their first discussions, one of the students had the idea of using yeast cells. If they could add a new little organ inside the cells, then this organ could be instructed to perform some useful task. First, they had to figure out how to defeat the cell's protection measures. Normally, a cell would send out a chemical "army" to destroy something new. The students used their genetic tools—the "scissors" and "glue" of genetic technology—to insert DNA into the yeast cell nucleus. They selected DNA that instructed the cell not to destroy the new organ they were creating. They called their new little organ a "synthesome."

The judges kept the contestants waiting for their decision. Finally, the high school team learned that it was a finalist, beating out college teams from major universities.

New Building Blocks of Science

In the 1800s, the new tools of mechanical engineering allowed scientists and engineers to build bridges and factories. In the 1900s, scientists used the tools of physics to invent the computer and put a man on the moon. Will the tools of genetic technology be the future of the 2000s? Many biologists believe so.

Today, biologists can sit at their computers and call up a list of hundreds of DNA strings. Each of these bits of DNA has been identified according to its use within a living organism. The idea of inserting a bit of DNA into the cell of a plant or animal has become almost commonplace. Do plant scientists want a blue rose? They look on their computers for a bit of DNA that makes a blue dye. Then, they set about inserting this DNA into the cells of a rose plant. Do fish farmers want to grow bigger salmon? They hire a geneticist to look for a gene that acts as a fish growth hormone. When this gene is inserted to make a transgenic salmon, the salmon puts on a growth spurt.

Synthetic biologists aim to take genetic engineering to the next higher step. They rearrange genes on an even larger scale. By attaching many bits of DNA together, it may be possible to put together the entire genome of a simple organism. Because the field is so new, scientists are only beginning to realize what may be possible. They seek to make artificial biological systems for engineering applications.

One possibility is that scientists could put together bits of DNA that would program a cell to produce energy. If this energy could be produced on a large scale, then it could power our homes

Botanists and plant biologists use genetic techniques to produce plants with improved features.

and automobiles. Someday soon, a genetic scientist might discover the equivalent of a Texas oil well while working at his or her laboratory bench. Now that the world is aware of the problems of air pollution and climate change that accompany the use of carbon fuels, scientists are looking for a new fuel source without these problems.

Just like the team of high school students in California, scientists in laboratories around the world are building new biological systems using genetic building blocks. The range of projects is almost unimaginable. For every human need, there is a good chance that a scientist somewhere is looking for a way to string DNA together to meet that need. These scientists include not only biologists but also engineers, computer scientists, mathematicians, chemists, and physicians. Helping them make the bridge to society are journalists, educators, ethicists, counselors, attorneys, and pharmacists.

Creating New Life?

In the words of genetics pioneer J. Craig Venter, as quoted in the *Guardian*, "We are going from reading our genetic code to the ability to write it." Dr. Venter is one of many scientists working to build an entire living cell, starting with basic chemical building blocks. These scientists want to build "living machines" designed to serve human purposes. Already scientists have been able to make an artificial genome from chemicals.

These "living machines" could be very useful to our society. Synthetic organisms may be able to produce gasoline substitutes for our cars, clean up pollution, or manufacture all kinds of medicines. But some scientists fear they could also bring unexpected risks. What happens if they escape from the laboratory into the environment? Could people be harmed in ways not anticipated? These scientists warn that there should be controls on new life forms and that society should proceed slowly in using them.

Others raise questions beyond the realm of science. Who will own the rights to these new life forms? Who will control how they are used? And what do these artificial forms of life say about what it means for a thing to be alive? With today's fast pace of science and exchange of information via the Internet, progress in the laboratory is rapid. Answering all the questions that are raised by the new science, however, is a much slower process.

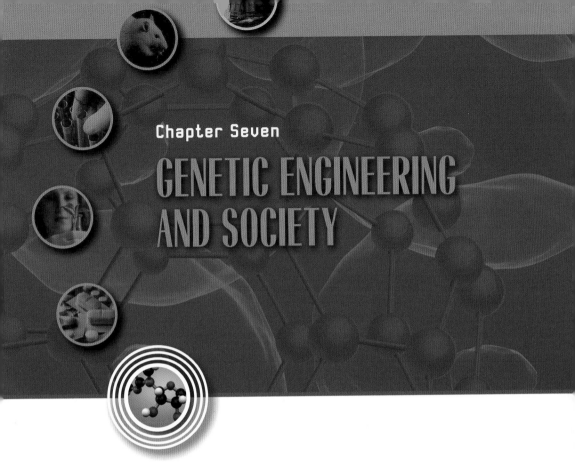

Chapter Seven

GENETIC ENGINEERING AND SOCIETY

J essie loves to play hockey. She is the goalie on her school's championship hockey team and the only girl on the team. She is also deaf. Some people think deaf babies and babies with other disabilities should never be born. They claim society would be better off with only "perfect" babies. Jessie would strongly disagree with them.

Scientists have discovered how to select babies and accomplish other amazing feats using what they've learned about genes. But should people take advantage of their new knowledge? If some new technology creates risks but also does a lot of good, then should it be used? Who decides if the benefits outweigh the risks? Who profits from the new technology? Can people believe all the exaggerated claims they hear? What if some people believe that using a procedure is right but others find it offensive to their beliefs? Because there are no easy answers, probably it's best for people to have an open mind and learn as much as they can about the issues.

Environment

Pollution seems to be everywhere—in the air, lakes and rivers, beaches, farm fields, and industrial sites. Once it's there, it's hard to clean up. Transgenic organisms may be able to break down harmful chemicals or clean up radiation spills. A transgenic pig produces waste that is less polluting to the environment.

Plant pathologists use genetic engineering to help farmers control weeds and pests. Genetic engineering also helped to produce crops with larger yields and plants with improved nutritive value.

What's for Dinner?

The answer to the question "What's for dinner?" is too often "Salmon." Numbers of wild salmon are down because of over-fishing. Fish suppliers have turned to genetic engineering to produce a fast-growing salmon they can raise in fish farms. But what happens if the genetically engineered salmon escape from their coastal fish farms and breed with wild populations? These salmon have a "super" gene that makes them grow quickly, and they might outcompete the wild salmon. Fish farmers claim that even if the salmon do escape their cages, they will not be able to breed with wild fish. Fish farms are regulated by government agencies that are responsible for both the quality of the food supply and environmental protection. But will there be enough safety checks to ensure that all the rules are followed by all fish farmers?

The Atlantic salmon at the top in this photo grew at a faster rate than its sibling, below, because of an added engineered gene for production of growth hormone. Both salmon are about eighteen months of age. Consequences of interbreeding between these "super" fish and native fish are unknown.

What about the harm that genetic engineering might cause, though? Transgenic plants and animals in the environment might interfere with the natural species that are already living there. In the 1990s, farmers started planting a transgenic form of corn. At first, it was thought that the corn was killing monarch butterflies, but this turned out not to be true. Still, scientists warn that such effects are possible with transgenic plants. Fish have escaped from fish farms and mixed with native ocean fish. If transgenic fish are raised in fish farms, then they could escape and pass on their altered genes to native fish.

Health and Safety

Many people already benefit from the medical uses of genetic engineering. But could genetic engineering also make people sick? Chances are you've already eaten genetically engineered food without even knowing it. According to the U.S. Food and Drug Administration, there is no reason for concern about eating genetically engineered foods. Others claim that food-testing methods are not adequate to assure food safety. Altered genes in food may lead to allergies, or cause resistance to antibiotics, or interfere with body chemicals.

Some people just do not like the idea of eating food from plants and animals that are genetically changed. They prefer to eat only natural foods. If food producers are required to label genetically modified food, then these people would have a choice. But requiring labels could make food more costly for everyone.

Accidents in genetics laboratories are a major safety concern. The accidental release of a genetically engineered disease "bug" might infect many people. Some fear that criminals might purposely engineer and release a disease bug. Genetic engineering laboratories are becoming common around the world. This accessibility increases the chance for an accident or criminal activity to occur. More safety controls for the labs may be needed.

WATCH OUT: LEGAL LOOPHOLE!

GMOs IN ANIMAL FEED

EU labelling rules contain a loophole: milk, eggs, meat and other animal-derived foodstuffs do not need to be labelled if the animals have been fed with GMOs. This loophole is directly responsible for GMOs continuing to be grown, and European consumers unwittingly supporting the cultivation of GMOs for animal feed. Labelling is essential to guarantee freedom of choice.

We demand mandatory labelling of animal products based on GMOs because ... a fundamental right in the European Union.

You can copy this list

European Union Commissioner for Consumer Protection Markos Kyprianou looks at a Greenpeace petition that calls for the labeling of genetically modified food. More than one million people in the European Union signed the petition calling for the labeling of food products derived from animals that are fed with genetically modified crops.

Privacy

Would you want to share your genetic code with the whole world? Your genetic code has information about what you are today and about your future health prospects. Although U.S. law protects citizens from discrimination by employers or health insurers, there may be other reasons for not sharing genetic information with employers, insurers, family, or friends. Sometimes, individuals are asked by medical professionals to contribute DNA to be part of a large group of DNA samples that are available for medical research. It is important that strict data confidentiality procedures are followed for these "genetic banks" so that the genetic privacy of individuals can be protected.

Economic and Social Issues

Many people hope to get rich from the applications of genetic engineering. Even if making money is not an issue, genetic research is expensive because of new technology. Medical treatments that involve genetic engineering are usually costly. It may not be possible for everyone to take advantage of these new medical options. Does this mean that the best medical care will go only to the rich? Many people regard this expense as unfair.

The companies that produce genetically modified seeds do not allow farmers to save the seeds to plant the next year. Instead, farmers must buy more seeds from the company. Some people say that genetically engineered plants and animals will provide more food to feed the hungry. Poor people worldwide will be better off. Others say this is untrue. They claim that the big companies will make money and small family farmers will suffer.

Religious and Moral Values

Does genetic engineering go against nature? Or is genetic engineering just another way humans can use their natural abilities to improve their world? There are no right or wrong answers to these questions. Most people decide that at least some forms of genetic engineering are acceptable, once they learn more about them.

Should plants and animals be genetically engineered for human use? Maybe it is all right to change a strawberry plant to make a sweeter berry. Is it all right to genetically engineer a goat to produce spider silk in her milk? Or to make a copy of a cow that produces a lot of beef? Researchers tell us that genetically engineered mice help them learn about human diseases.

Many people believe that it is good to use plants and animals if doing so helps humans. In the future, doctors may be able to cure dreaded diseases, using what has been learned from genetic engineering of plants and animals. Everyone agrees that animals should be treated well and that there should be limits on how animals are used in science. But some people think it is never right for humans to change the natural genetics of animals. They say it is more important to follow the laws of nature than to perform such medical research.

The most disturbing issues are those involving human genetic engineering. Almost everyone agrees it is unethical to clone a human being. People are used to the idea that they are different from one another. A cloned human being would not have the chance to be his or her own special self.

Depending on a person's beliefs, genetic testing may be acceptable to assure the birth of a healthy baby. But what if parents used these methods to be sure they had a baby girl? Some decisions are very personal, and each individual has to consider such personal issues for him- or herself.

Hype or Fact—What to Believe?

Many individuals and organizations have strong beliefs about genetic engineering. Sometimes, these strong beliefs lead to exaggeration. Those who favor a certain technique may make sweeping statements about how much good it can do. However, those opposed to it may overstate the harm that can come from it. For an individual trying to decide about these issues, it's hard to be sure just what to believe. Extreme statements on either side are unlikely to be true. Most likely the truth lies somewhere in the middle.

It is exciting to be a part of this new Age of Biology. Procedures and treatments never before thought possible happen almost daily. Cures for dreaded diseases may be just around the corner. A new fuel source may head off climate change. Each new step, though, brings more questions. Everyone stands to gain from the future of genetic engineering. Likewise, everyone shares in its risks. Somehow, people worldwide must decide how the benefits of genetic engineering can be shared fairly and how to protect everyone from the risks. The best decisions will come from well-educated citizens who understand the science, the benefits, and the drawbacks of genetic engineering and its applications.

GLOSSARY

bacteria One-celled organisms that divide rapidly. Genetically engineered bacteria can produce medicines and other essential chemicals.

biotechnology Methods and techniques involving use of living organisms to produce chemicals or provide for other human needs.

chromosomes Structures within a cell that contain genetic material in the form of long strands of DNA.

clone A genetically identical copy of an organism.

designer babies Babies with genetic material selected or altered prior to birth, using genetic engineering, to assure the presence or absence of some trait.

DNA (deoxyribonucleic acid) A complex chemical found in all cells that contains the organism's genetic code. The chemical has the shape of a double helix.

embryo A stage in the process of reproduction that lasts, in humans, from about ten days after egg fertilization up to about eight weeks.

gene The physical unit of inheritance within a cell. It consists of a sequence of chemical bases at a particular location on a DNA molecule.

gene therapy A process of treating disease that involves altering a person's genetic material.

genetic disease A disease or disorder that is inherited genetically and is due to faulty genetic material.

genetic testing Laboratory analysis of a tissue sample to determine if a person's genetic material contains a particular gene or genes.

genome All the DNA in an organism's complete set of chromosomes.

in vitro fertilization A medical procedure in which an egg is removed from the ovary and fertilized in a laboratory environment, and the resulting embryo is placed in the uterus.

mutation A permanent change in the DNA of a gene. Mutations can be inherited or can occur in individual cells throughout an organism's lifetime.

nucleus A structure in a cell within which chromosomes are located.

penetrance The chance that having a particular gene will actually result in disease. If a gene has high penetrance, then it is very likely a person whose genetic code contains that gene will get the disease.

recombinant DNA DNA that is produced by combining DNA from different sources.

restriction enzyme A chemical used in genetic engineering to cut a strand of DNA at a particular sequence of bases.

stem cell A cell that can divide almost indefinitely and that is capable of producing various types of specialized cells.

synthetic biology The building of new biological systems from parts. It uses genetic engineering methods to put together new DNA from chemicals or selected DNA sequences.

transgenic animal An animal that has had genetic material from another species inserted into its DNA prior to birth.

virus A tiny microorganism that depends on another cell for growth and reproduction. It is useful in genetic engineering because of its ability to enter a cell and interfere with the cell's DNA.

FOR MORE INFORMATION

Canadian Genetic Diseases Network
Networks of Centres of Excellence
350 Albert Street, 16th Floor Mailroom
Ottawa, ON K1A 1H5
Canada
(613) 995-6010
Web site: http://www.nce.gc.ca/nces-rces/cgdn_e.htm
The network's mission is to advance genetics research and the
    application of genetic discoveries to the prevention, diagnosis,
    and treatment of human disease.

Genome Canada
150 Metcalfe Street, Suite 2100
Ottawa, ON K2P 1P1
Canada
(613) 751-4460
Web site: http://www.genomecanada.ca
This funding and information resource is dedicated to promoting
    genomics research to benefit all Canadians.

Human Genome Project
U.S. Department of Energy
1000 Independence Avenue SW
Washington, DC 20585
(800) DIAL-DOE (342-5363)
Web site: http://www.ornl.gov/sci/techresources/Human_
    Genome/home.shtml#index
The goals of this project are to identify the entire sequence of the
    human genome, to make this information publicly available,
    and to address related ethical issues.

The Humane Society of the United States
2100 L Street NW
Washington, DC 20037
(202) 452-1100
Web site: http://www.hsus.org
This society works to prevent cruelty, exploitation, and neglect
of animals.

National Institutes of Health (NIH)
9000 Rockville Pike
Bethesda, MD 20892
(301) 496-4000
Web site: http://www.nih.gov
The NIH is the research and medical agency of the United
States. Its mission is to provide knowledge about science and
the behavior of living systems, and the application of that
knowledge for living a healthy life and for reducing illness
and disabilities.

National Office of Public Health Genomics
Centers for Disease Control and Prevention
4770 Buford Highway Mailstop K-89
Atlanta, GA 30341
(800) 488-8510
Web site: http://www.cdc.gov/genomics/default.htm
This U.S. government agency promotes the integration of genomics into public health research, policy, and practice in order
to improve the lives and health of all people.

Union of Concerned Scientists
2 Brattle Square
Cambridge, MA 02238
(617) 547-5552
Web site: http://www.ucsusa.org

This science-based alliance of scientists and citizens combines scientific research and citizen action to promote responsible changes in government policy and industry practices as they relate to science issues.

U.S. Food and Drug Administration (FDA)
5600 Fishers Lane
Rockville, MD 20857-0001
(888) INFO-FDA (463-6332)
Web site: http://www.fda.gov
The mission of the FDA is to assure the safety of the nation's food supply and drugs, to promote innovations in food and drug production, and to educate the public.

Web Sites

Due to the changing nature of Internet links, Rosen Publishing has developed an online list of Web sites related to the subject of this book. This site is updated regularly. Please use this link to access the list:

http://www.rosenlinks.com/sas/mgs

FOR FURTHER READING

Bickerstaff, Linda. *Technology and Infertility: Assisted Reproduction and Modern Society*. New York, NY: The Rosen Publishing Group, Inc., 2009.

Boskey, Elizabeth. *America Debates Genetic DNA Testing*. New York, NY: The Rosen Publishing Group, Inc., 2008.

Cefrey, Holly. *Cloning and Genetic Engineering*. New York, NY: Rosen Book Works, Inc., 2002.

Freedman, Jeri. *Genetically Modified Food: How Biotechnology Is Changing What We Eat*. New York, NY: The Rosen Publishing Group, Inc., 2005.

Freedman, Jeri. *How Do We Know About Genetics and Heredity*. New York, NY: The Rosen Publishing Group, Inc., 2005.

Fridell, Ron. *Genetic Engineering*. Minneapolis, MN: Lerner Publications Company, 2006.

Grace, Eric S. *Biotechnology Unzipped: Promises and Realities*. 2nd ed. Washington, DC: Joseph Henry Press, 2006.

Stoyles, Pennie, Peter Pentland, and David Demant. *Science Issues: Genetic Engineering*. North Mankato, MN: Smart Apple Media, 2003.

BIBLIOGRAPHY

Avise, John C. *The Hope, Hype, and Reality of Genetic Engineering*. New York, NY: Oxford University Press, 2004.

Crichton, Michael. *Next*. New York, NY: HarperCollins Publishers, 2006.

Dyson, Freeman. "Our Biotech Future." *New York Review of Books*, July 19, 2007. Retrieved January 31, 2008 (http://www.nybooks.com/articles/20370).

Glenn, Linda M. "Ethical Issues in Genetic Engineering and Transgenics." ActionBioscience.org, June 2004. Retrieved January 31, 2008 (http://www.actionbioscience.org/biotech/glenn.html).

Jurassic Park (movie). Universal City, CA: Universal Studios, 1993.

Kolata, Gina. "Scientists Bypass Need for Embryo to Get Stem Cells." *New York Times*, November 21, 2007. Retrieved February 7, 2008 (http://www.nytimes.com/2007/11/21/science/21stem.html?scp=49&sq=stem+cell&st=nyt).

Korn, Peter. "Groups Push a Genetic Bank." *Portland Tribune*, April 11, 2008. Retrieved May 28, 2008 (http://www.portlandtribune.com/news/story.php?story_id=120785920961939800).

Marjadi, Meghna. "Genome Mapping for Fun and Profit." *McGill Tribune*, January 8, 2008. Retrieved January 31, 2008 (http://media.www.mcgilltribune.com/media/storage/paper234/news/2008/01/08/Features/Genome.Mapping.For.Fun.And.Profit-3147032.shtml).

National Human Genome Research Institute. "Genetic Information Nondiscrimination Act of 2008." Retrieved May 28, 2008 (http://www.genome.gov/1002328).

Pollack, Andrew. "Engineering by Scientists on Embryo Stirs Criticism." *New York Times*, May 13, 2008. Retrieved

May 13, 2008 (http://www.nytimes.com/2008/05/13/
science/13embryo.html?_r=1&...).

Sandel, Michael J. *The Case Against Perfection: Ethics in the
Age of Genetic Engineering*. Cambridge, MA: Belknap Press
of Harvard University Press, 2007.

Silver, Lee M. *Challenging Nature: The Clash Between
Biotechnology and Spirituality*. New York, NY: HarperCollins
Publishers, 2006.

Tansey, Bernadette. "High School Biowizards Break New Ground
in Winning Competition." SFGate.com, November 17, 2007.
Retrieved January 31, 2008 (http://www.sfgate.com/cgi-bin/
article.cgi?f=/c/a/2007/11/17/MN4KTCC44.DTL).

Wade, Nicholas. "Genetic Engineers Who Don't Just Tinker."
New York Times, July 8, 2007. Retrieved January 31, 2008
(http://www.nytimes.com/2007/07/08/weekinreview/
08wade.html?_r=3&ref=science&oref=slogin&oref=slogin&
oref=slogin).

INDEX

About the Author

Terry L. Smith, MS, is a statistician and science writer who resides in Lawrence, Kansas. She formerly served on the faculty of the University of Texas M.D. Anderson Cancer Center in Houston. Smith is the author of numerous research articles and science books, including her most recent book, *Frequently Asked Questions About Celiac Disease*. Her continuing interest in genetic engineering stems from her professional involvement in clinical trials of several genetically engineered drugs for the treatment of cancer.

Photo Credits

Cover © Keith Weller/U.S. Department of Agriculture/Photo Researchers; cover (inset) © www.istockphoto.com/Chris Dascher; p. 5 © MCA Universal Pictures/Photofest; p. 6 © www.istockphoto.com; p. 9 © Dr. K. G. Murti/Visuals Unlimited; p. 10 © Maggie Caldwell Gill/Custom Medical Stock Photo; p. 12 © IMS Creative/Custom Medical Stock Photo; pp. 14, 25 © Terese Winslow; p. 16 © James King-Holmes/Photo Researchers; p. 17 © Getty Images; pp. 18, 49 © AFP/Getty Images; p. 21 © Michael Newman/Photo Edit; p. 22 © Phanie/Photo Researchers; p. 23 © David Wrobel/Visuals Unlimited; p. 28 © www.istockphoto.com/Brian Swartz; p. 30 © AP Photo; p. 29 www.glofish.com; p. 35 © Hop Americain/Photo Researchers; p. 36 © Anatomical Travelogue/Photo Researchers; p. 40 Abraham Lincoln High School; p. 42 © Scott Bauer/U.S. Department of Agriculture/Photo Researchers; p. 46 U.S. Department of Agriculture; p. 47 © 2002 Aqua Bounty.

Designer: Evelyn Horovicz; Cover Designer: Nelson Sá
Editor: Kathy Kuhtz Campbell; Photo Researcher: Marty Levick